Learn

Eureka Math™
Grade K
Module 3

Published by Great Minds®.

Copyright © 2018 Great Minds®.

Printed in the U.S.A.

This book may be purchased from the publisher at eureka-math.org.

10   9   8   7   6   5   4   3   2   1

ISBN 978-1-64054-077-4

GK-M3-L-05.2018

Learn • Practice • Succeed

Eureka Math™ student materials for *A Story of Units*® (K–5) are available in the *Learn, Practice, Succeed* trio. This series supports differentiation and remediation while keeping student materials organized and accessible. Educators will find that the *Learn, Practice,* and *Succeed* series also offers coherent—and therefore, more effective—resources for Response to Intervention (RTI), extra practice, and summer learning.

Learn

Eureka Math Learn serves as a student's in-class companion where they show their thinking, share what they know, and watch their knowledge build every day. *Learn* assembles the daily classwork—Application Problems, Exit Tickets, Problem Sets, templates—in an easily stored and navigated volume.

Practice

Each *Eureka Math* lesson begins with a series of energetic, joyous fluency activities, including those found in *Eureka Math Practice.* Students who are fluent in their math facts can master more material more deeply. With *Practice,* students build competence in newly acquired skills and reinforce previous learning in preparation for the next lesson.

Together, *Learn* and *Practice* provide all the print materials students will use for their core math instruction.

Succeed

Eureka Math Succeed enables students to work individually toward mastery. These additional problem sets align lesson by lesson with classroom instruction, making them ideal for use as homework or extra practice. Each problem set is accompanied by a Homework Helper, a set of worked examples that illustrate how to solve similar problems.

Teachers and tutors can use *Succeed* books from prior grade levels as curriculum-consistent tools for filling gaps in foundational knowledge. Students will thrive and progress more quickly as familiar models facilitate connections to their current grade-level content.

Students, families, and educators:

Thank you for being part of the *Eureka Math*™ community, where we celebrate the joy, wonder, and thrill of mathematics.

In the *Eureka Math* classroom, new learning is activated through rich experiences and dialogue. The *Learn* book puts in each student's hands the prompts and problem sequences they need to express and consolidate their learning in class.

What is in the Learn *book?*

Application Problems: Problem solving in a real-world context is a daily part of *Eureka Math*. Students build confidence and perseverance as they apply their knowledge in new and varied situations. The curriculum encourages students to use the RDW process—Read the problem, Draw to make sense of the problem, and Write an equation and a solution. Teachers facilitate as students share their work and explain their solution strategies to one another.

Problem Sets: A carefully sequenced Problem Set provides an in-class opportunity for independent work, with multiple entry points for differentiation. Teachers can use the Preparation and Customization process to select "Must Do" problems for each student. Some students will complete more problems than others; what is important is that all students have a 10-minute period to immediately exercise what they've learned, with light support from their teacher.

Students bring the Problem Set with them to the culminating point of each lesson: the Student Debrief. Here, students reflect with their peers and their teacher, articulating and consolidating what they wondered, noticed, and learned that day.

Exit Tickets: Students show their teacher what they know through their work on the daily Exit Ticket. This check for understanding provides the teacher with valuable real-time evidence of the efficacy of that day's instruction, giving critical insight into where to focus next.

Templates: From time to time, the Application Problem, Problem Set, or other classroom activity requires that students have their own copy of a picture, reusable model, or data set. Each of these templates is provided with the first lesson that requires it.

Where can I learn more about Eureka Math *resources?*

The Great Minds® team is committed to supporting students, families, and educators with an ever-growing library of resources, available at eureka-math.org. The website also offers inspiring stories of success in the *Eureka Math* community. Share your insights and accomplishments with fellow users by becoming a *Eureka Math* Champion.

Best wishes for a year filled with aha moments!

Jill Diniz

Jill Diniz
Director of Mathematics
Great Minds

The Read–Draw–Write Process

The *Eureka Math* curriculum supports students as they problem-solve by using a simple, repeatable process introduced by the teacher. The Read–Draw–Write (RDW) process calls for students to

1. Read the problem.

2. Draw and label.

3. Write an equation.

4. Write a word sentence (statement).

Educators are encouraged to scaffold the process by interjecting questions such as

- What do you see?

- Can you draw something?

- What conclusions can you make from your drawing?

The more students participate in reasoning through problems with this systematic, open approach, the more they internalize the thought process and apply it instinctively for years to come.

Contents

Module 3: Comparison of Length, Weight, Capacity, and Numbers to 10

Topic F: Comparison of Sets Within 10

Topic G: Comparison of Numerals

Topic H: Clarification of Measurable Attributes

Draw a skyscraper.

Draw a one-story building.

 Draw

(Show students a picture of a skyscraper and a one-story building.) With your partner, describe how the skyscraper and the one-story building are the same and how they are different. (Model several comparative sentences using student descriptions of the buildings.)

Lesson 1: Compare lengths using *taller than* and *shorter than* with aligned and non-aligned endpoints.

©2018 Great Minds®. eureka-math.org

1

Name _____   Date _____

For each pair, circle the longer one. Imagine the paper strips are lying flat on a table.

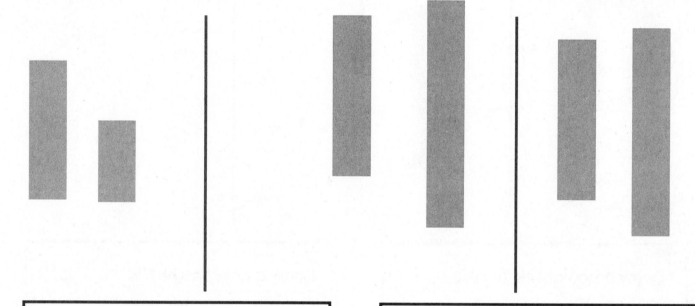

Draw a flower that is taller than the vase.

Draw a tree that is taller than the house.

For each pair, circle the shorter one.

Draw a bookmark that is shorter than this book.

Draw a crayon that is shorter than this pencil.

Lesson 1:   Compare lengths using *taller than* and *shorter than* with aligned and non-aligned endpoints.

EUREKA MATH™

Draw something that is very tall.

Draw

[blank drawing box]

Compare your picture of something tall with a friend's. Is the item in your friend's drawing shorter or taller than the item in your drawing? How do you know?

Name _____   Date _____

Cut out the picture of the string at the bottom of the page. Compare the string with each object to see which is longer. Use the line next to each object to help you compare. Color objects shorter than the string green. Color objects longer than the string orange.

On the back of your paper, draw something longer than, shorter than, and the same length as the picture of the string. Color objects shorter than the string green. Color objects longer than the string orange.

Longer or Shorter Recording Sheet

These objects are **longer than** my string.	These objects are **shorter than** my string.

_____

longer or shorter

Draw a monkey with a very long tail.

Draw a monkey with a very short tail.

Draw a banana for the monkeys to share.

Draw

Compare the banana to the tails of both monkeys. Is the banana longer or shorter than the long tail? Is the banana longer or shorter than the short tail?

Directions: Pretend that I am a pirate who has traveled far away from home. I miss my house and family. Will you draw a picture as I describe my home? Listen carefully, and draw what you hear.

- Draw a house in the middle of the paper as tall as your pointer finger.

- Now, draw my daughter. She is shorter than the house.

- There's a great tree in my yard. My daughter and I love to climb the tree. The tree is taller than my house.

- My daughter planted a beautiful daisy in the yard. Draw a daisy that is shorter than my daughter.

- Draw a branch lying on the ground in front of the house. Make it the same length as the house.

- Draw a caterpillar next to the branch. My parrot loves to eat caterpillars. Of course, the length of the caterpillar is shorter than the length of the branch.

- My parrot is always hungry, and there are plenty of bugs for him to eat at home. Draw a ladybug above the caterpillar. Should the ladybug be shorter or longer than the branch?

- Now, draw some more things you think my family would enjoy.

Show your picture to your partner, and talk about the extra things that you drew. Use *longer than* and *shorter than* when you are describing them.

Name _____ Date _____

Home is where the heARRt is, matey.

Lesson 3: Make a series of *longer than* and *shorter than* comparisons.

Longer than...

Shorter than...

---

longer than and shorter than work mat

Draw two things that make the sentences true.

Say the sentences to your partner.

Draw

(Write these sentence frames on the board and read them aloud to the class.)

I am taller than _____.   I am shorter than _____.

(Listen as students say each sentence out loud to their partner. Provide support as needed.)

Name _____   Date _____

Circle the shorter stick.

How many linking cubes are in the shorter stick? Write the number in the box.

How many linking cubes are in the shorter stick? Write the number in the box.

Circle the longer stick.

How many linking cubes are in the longer stick? Write the number in the box.

How many linking cubes are in the longer stick? Write the number in the box.

Lesson 4:   Compare the length of linking cube sticks to a 5-stick.

19

©2018 Great Minds®. eureka-math.org

Draw a stick **shorter than** my 5-stick.

Draw a stick **longer than** mine.

Draw a stick **shorter than** mine.

Longer than my 5-stick:

Shorter than my 5-stick:

longer or shorter mat

Lesson 4:   Compare the length of linking cube sticks to a 5-stick.

21

Write your first name so that one letter is in each box.

Start at the box above the star. Don't skip any boxes.

 **Draw**

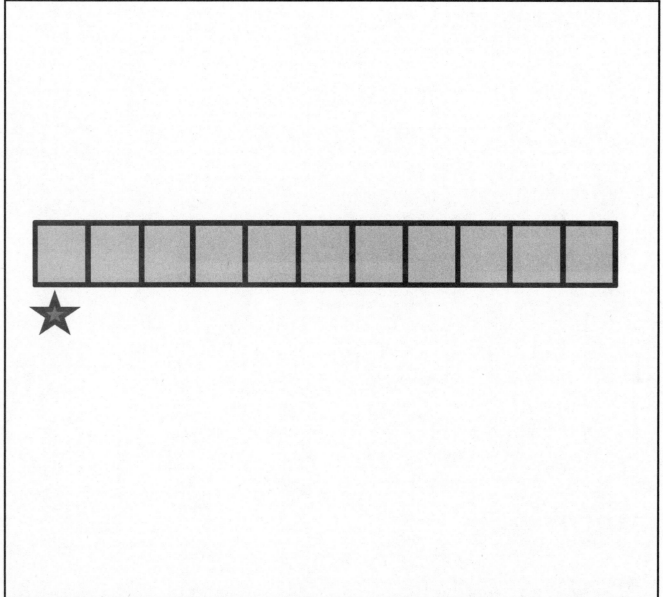

You made a name train. Compare your name train to your partner's. Which train has fewer letter passengers? Which train has more letter passengers? (Listen to comparative statements and observe endpoint alignment skills.)

Lesson 5:   Determine which linking cube stick is *longer than* or *shorter than* the other.

©2018 Great Minds®. eureka-math.org

23

Name _____   Date _____

Circle the stick that is shorter than the other.

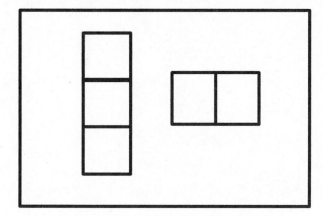 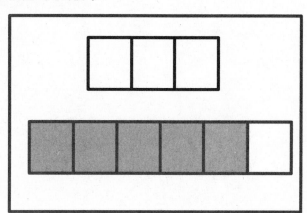

Circle the stick that is longer than the other.

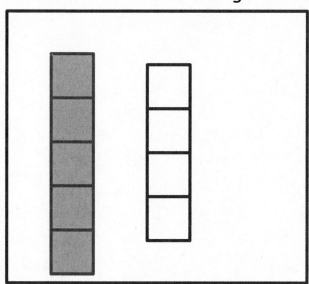

My _____ -stick is longer than my _____ -stick.

My _____ -stick is shorter than my _____ -stick.

Lesson 5:   Determine which linking cube stick is *longer than* or *shorter than* the other.

©2018 Great Minds®. eureka-math.org

25

Circle the stick that is shorter than the other stick.

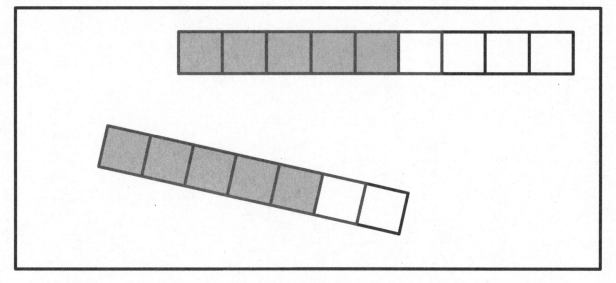

My _____ -stick is longer than my _____ -stick.

My _____ -stick is shorter than my _____ -stick.

On the back of your paper, draw a 6-stick.

Draw a stick longer than your 6-stick.

Draw a stick shorter than your 6-stick.

OR

On the back of your paper, draw a 9-stick.

Draw a stick longer than your 9-stick.

Draw a stick shorter than your 9-stick.

Lesson 5:   Determine which linking cube stick is *longer than* or *shorter than* the other.

Put your hand on the paper and spread out your fingers.

Trace around your hand to make your handprint.

Draw

Look in the bag of linking cube stairs and find the one that is closest to the length of your thumb. Pull it out and check your guess by placing it on the thumb of your hand drawing. (Repeat with other finger lengths.)

Name _____   Date _____

In the box, write the number of cubes there are in the pictured stick.
Draw a green circle around the stick if it is longer than the object.
Draw a blue circle around the stick if it is shorter than the object.

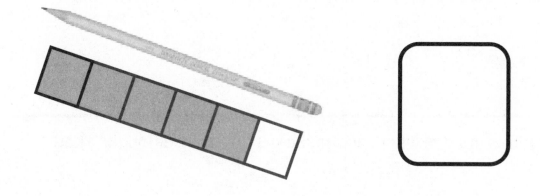

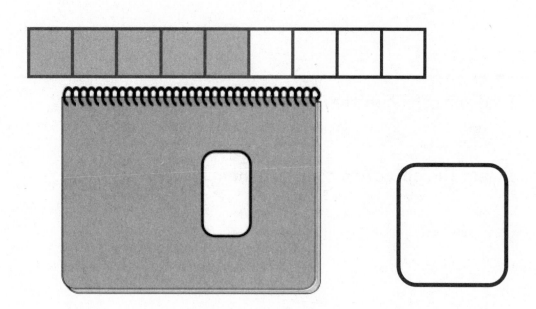

EUREKA
MATH™

Lesson 6:   Compare the length of linking cube sticks to various objects.

31

©2018 Great Minds®. eureka-math.org

Make a 3-stick. In your classroom, select a crayon, and see if your crayon is longer than or shorter than your stick.

Trace your 3-stick and your crayon to compare their lengths.

In your classroom, find a marker, and make a stick that is longer than your marker.

Trace your stick and your marker to compare their lengths.

Make a 5-stick. Find something in the classroom that is longer than your 5-stick.

Trace your 5-stick and the object to compare their lengths.

Lesson 6:   Compare the length of linking cube sticks to various objects.

EUREKA MATH

Make a clay snake as long as your pointer finger.

Make a clay snake as long as your pinky finger.

Draw or trace your snakes.

Draw

Circle the snake that is longer. Say a sentence comparing both of your snakes.

Name _____   Date _____

These boxes represent cubes.

Color 2 cubes red. Color 3 cubes green.

How many cubes did you color? ☐

Is this stick the same length as the gray stick?   YES   NO

Together 2 cubes and 3 cubes are the same length as 5.

Color 1 cube red and the rest green.

How many cubes did you color? ☐

Is this stick the same length as the gray stick?   YES   NO

Together 1 cube and 4 cubes are the same length as _____.

Trace a 6-stick. Find something the same length as your 6-stick.

Draw a picture of it here.

Trace a 7-stick. Find something the same length as your 7-stick.

Draw a picture of it here.

Trace an 8-stick. Find something the same length as your 8-stick.

Draw a picture of it here.

Lesson 7:   Compare objects using *the same as*.

EUREKA
MATH™

My 5:

My _____:

My _____:

_____

riddle work mat

My 5:

My _____:

My _____:

_____

riddle work mat

Lesson 7:   Compare objects using *the same as*.

EUREKA MATH

Draw 3 things you could easily carry in your backpack and not get tired.

Draw 1 thing that would make you tired if carried in your backpack.

**Draw**

3 things

1 thing

Talk to your partner about why carrying 1 thing could make you more tired than carrying 3 things.

Name _____   Date _____

Which is heavier? Circle the object that is heavier than the other.

On the back, draw 3 objects that are lighter than your chair.

Draw two things that make the sentence true.

Say the sentence to your partner.

Draw

```
┌─────────────────────────────────────────────────┐
│                                                 │
│                                                 │
│                                                 │
│                                                 │
│                                                 │
│                                                 │
│                                                 │
│                                                 │
│                                                 │
│                                                 │
│                                                 │
│                                                 │
│                                                 │
│                                                 │
│                                                 │
│                                                 │
└─────────────────────────────────────────────────┘
```

(Write the following sentence frame on the board and read it aloud to the class.) I am lighter than _____, but I am heavier than _____. (Listen as students say the sentence out loud. Provide support as needed.) How much do you think you weigh?

Lesson 9:   Compare objects using *heavier than, lighter than,* and *the same as* with balance scales.

43

©2018 Great Minds®. eureka-math.org

Name _____   Date _____

Lighter   Heavier

_____

lighter or heavier recording sheet

EUREKA MATH   **Lesson 9:**   Compare objects using *heavier than, lighter than,* and *the same as*   **45**
with balance scales.

©2018 Great Minds®. eureka-math.org

Imagine you were on a seesaw with a little kitten on the other end.

Draw which end you would be on and which end the kitten would be on.

Draw

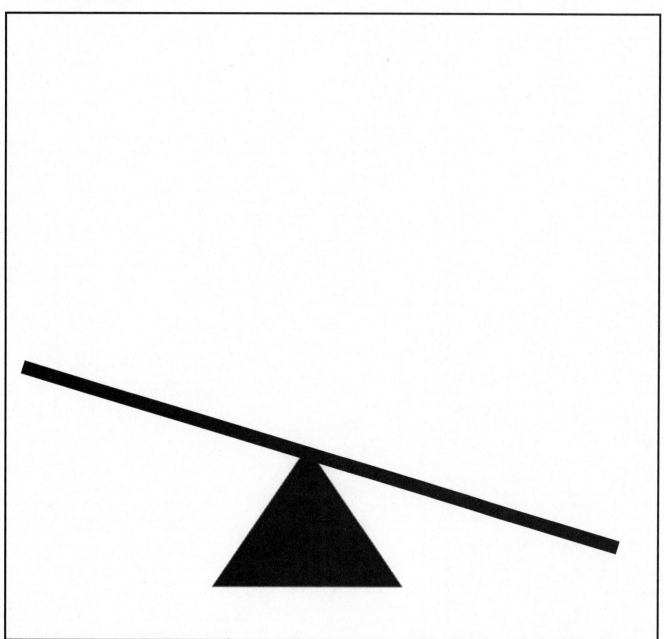

Are you on the end closer to the ground? Why or why not? Where is the kitten? Tell your partner why you drew the kitten where you did.

EUREKA MATH

Lesson 10: Compare the weight of an object to a set of unit weights on a balance scale.

47

©2018 Great Minds®. eureka-math.org

Lesson 10:   Compare the weight of an object to a set of unit weights on a balance scale.

Name _____   Date _____

is as heavy as _____ pennies.

is as heavy as _____ pennies.

is as heavy as _____ pennies.

is as heavy as _____ pennies.

as heavy as recording sheet

 Lesson 10:   Compare the weight of an object to a set of unit weights on a balance
scale.

49

Make a heavy building with your connecting blocks.

Use a balance scale to see how many pennies are as heavy as your building.

Draw your building and pennies on the balance scale.

 Draw

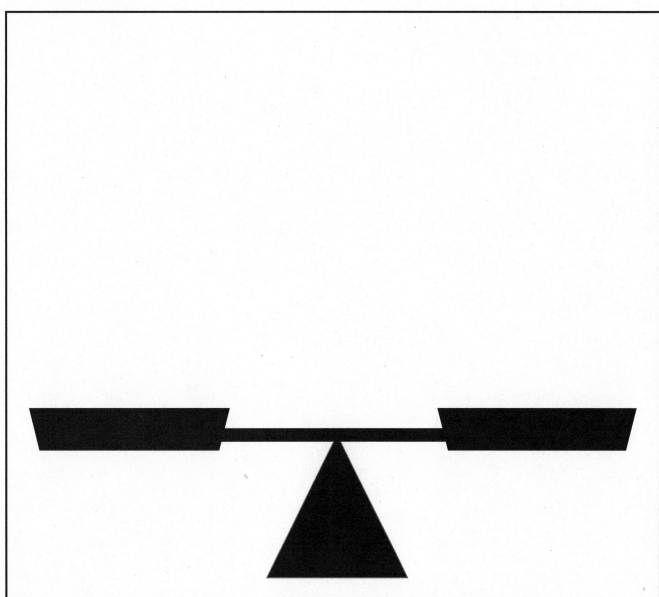

(Students make "heavy" buildings from small connecting blocks. Read the directions above and assist with the balance weighing when needed.)

Name _____   Date _____

Draw a line from the balance to the linking cubes that weigh the same.

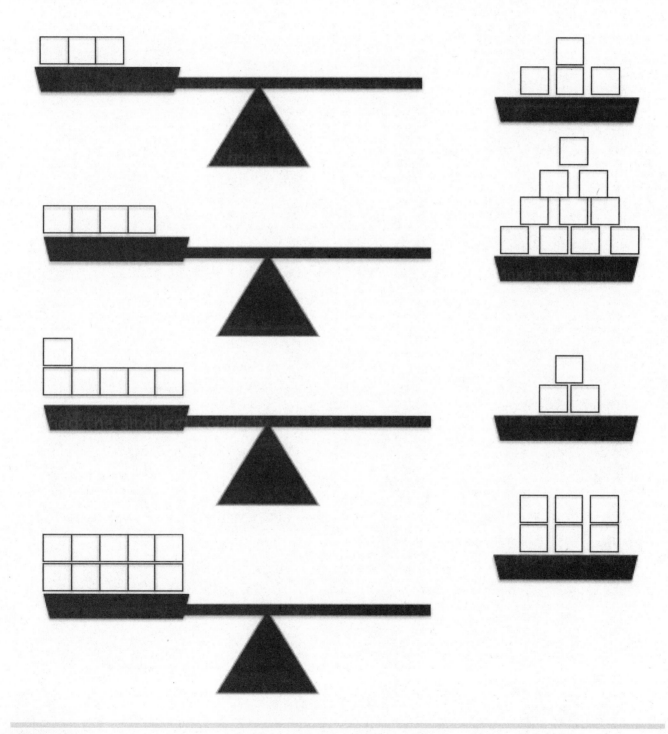

Put a small object on a balance scale.

How many pennies do you think it will take to balance your object?

Test your guess.

Draw your object and pennies on the balance scale.

Draw

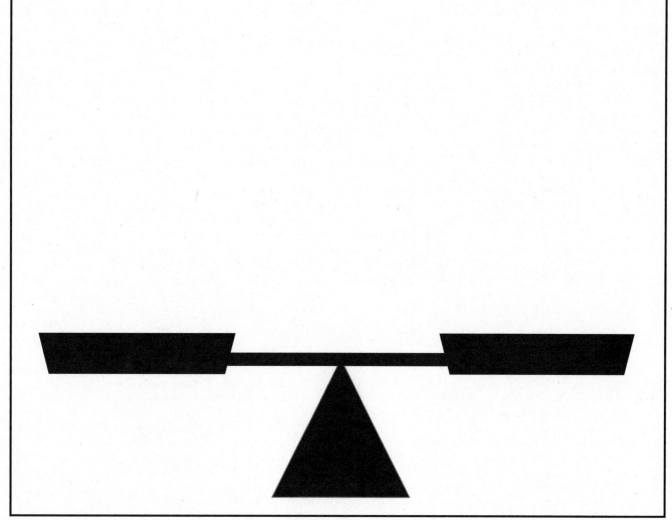

(Put the following sentence frame on the board.) My _____ is as heavy as a set of ____ pennies.
(Students say the comparative sentence.) Put another penny on both sides of the scale.
What happened?

Name _____   Date _____

My [   ]   is as heavy as a set of [ ]

My [   ]   is as heavy as a set of [ ]

My [   ]   is as heavy as a set of [ ]

My [   ]   is as heavy as a set of [ ]

_____
as heavy as a set recording sheet

With your clay, make a small cup that could hold just enough milk for a little kitten to drink.

Draw your cup.

Draw

Show your cup to a friend. Will your cups hold the same amount of milk?

Lesson 13:   Compare volume using *more than, less than,* and *the same as* by pouring.

©2018 Great Minds®. eureka-math.org

59

Name _____ Date _____

Talk to your partner about which container might have more or less capacity. Which might have about the same capacity? What happens if the containers are not filled up to the top? Can we tell that they are filled completely from looking at the pictures?

Lesson 13:   Compare volume using *more than, less than,* and *the same as* by pouring.

©2018 Great Minds®. eureka-math.org

61

Name _____   Date _____

I found out that this container held the most rice.

It had the biggest capacity.

I found out that this container held the least rice.

It had the smallest capacity.

capacity recording sheet

   Lesson 13:   Compare volume using *more than, less than,* and *the same as* by   63
pouring.

©2018 Great Minds®. eureka-math.org

With your clay, make a bowl big enough to hold a strawberry.

With your clay, make a little vase to hold a tiny flower.

Draw the bowl and vase.

Draw

Compare your containers. Write an *M* on the drawing that would hold more. Write an *L* on the drawing that would hold less.

Name _____   Date _____

My cup of rice looks like:

Now it looks like:

Now it looks like:

Now it looks like:

volume recording sheet

With your clay, make a container just big enough to hold 10 beans.

Put 10 beans in your container. Draw your container with the beans.

Draw

Is there room for more beans in your container? How many more beans do you think will fit?

Was your container too small? How many beans did not fit in your container?

Did your container hold 10 beans exactly with no room left over?

Name _____   Date _____

We've Got the Scoop!

_____ is the same as _____ scoops.

_____ is the same as _____ scoops.

_____ is the same as _____ scoops.

_____ scoops is the same as

_____ scoops is the same as
_____

we've got the scoop recording sheet

Guess how many linking cubes it would take to cover your card.

Use your linking cubes to cover your card and test your guess.

Write the number of cubes you used on the card shown.

Draw

(Give each student a playing card and linking cubes.) How many linking cubes did you use to cover your card? Did your friends use the same number of cubes?

Name _____   Date _____

Cover the shape with squares. Count how many, and write the number in the box.

Squares

Cover the shape with beans. Count how many, and write the number in the box.

Beans

Name _____   Date _____

My square.

My square covered
with a circle.

My square covered
with little squares.

My square covered
with beans.

_____
my square recording sheet

Draw enough chairs for each smiling face.

How many chairs did you draw?

Draw

(Play a round of musical chairs with several more chairs than students. Eliminate one chair each time the music stops until there are enough chairs for each student. Students draw just enough chairs on this sheet for the smiling faces to end the activity.)

Name _____   Date _____

Draw straight lines with your ruler to see if there are enough flowers for the butterflies.

On the back, draw some plates. Draw enough apples so each plate has one.

Draw 4 little mice.

Draw pieces of cheese so that each mouse can have one.

 Draw

Draw a line from each mouse to its cheese. Are there just enough pieces of cheese? Talk to your partner about how you knew how many pieces of cheese to draw.

Name _____   Date _____

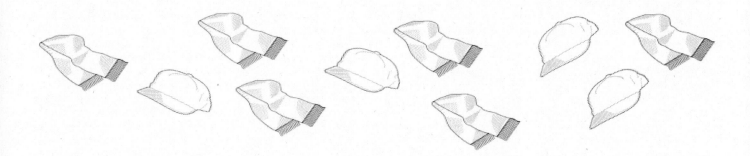

Draw straight lines with your ruler to see if there are enough hats for the scarves.

Are there more or [scarf] ?

Cross off by putting an X on 2 [scarf] . Talk to your partner about what you notice now.

Draw more leaves than ants.

Make 6 pancakes with your clay. How many people could you serve?
One more person wants a pancake. Put your clay back together and make
enough pancakes for everyone. Draw the pancakes.

Draw

What happened to the size of the pancakes when you had to make one more?

Name _____   Date _____

Count the objects. Circle the set that has fewer.

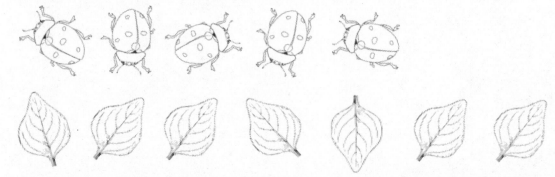

Draw more ladybugs so there are the same number of ladybugs as leaves.

Count the objects. Circle the set that has fewer.

Draw more watermelon slices so there are the same number of watermelon slices as peaches.

On the back, draw suns and stars. Draw fewer suns than stars.

Write your first name so that one letter is in each top box.

Then, write your last name below so that one letter is in each box.

Start at the box above the star. Don't skip any boxes.

Draw

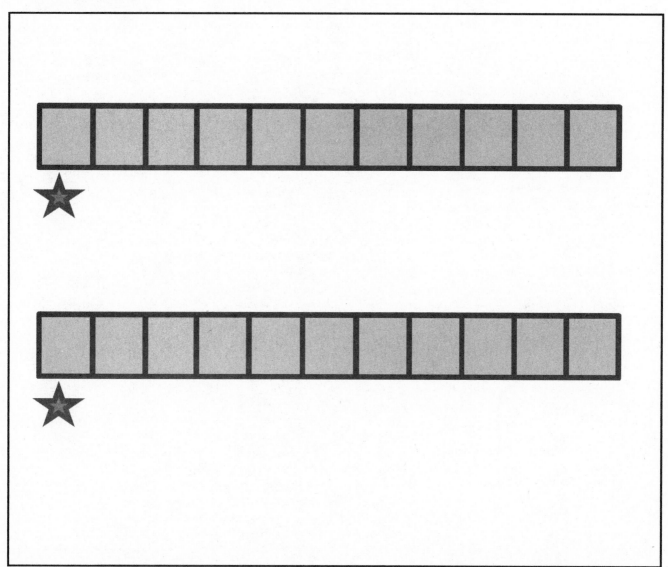

Which of your trains has more letter passengers? Which passenger train is longer? Which of your trains has fewer passengers? Which passenger train is shorter? Talk about your trains with your partner. Are your partner's trains similar to yours? Did anyone's train not have enough room for all of the letter passengers?

Lesson 20:   Relate *more* and *less* to length.

Name _____   Date _____

Count the dots on the die. Color as many beads as the dots on the die.
Circle the longer chain in each pair.

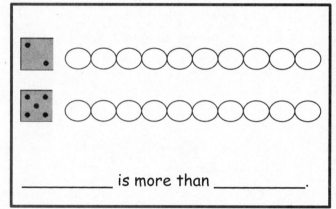

_____ is more than _____.

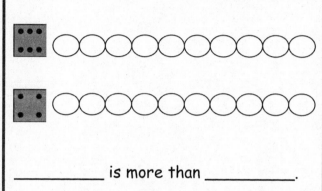

_____ is more than _____.

Roll the die. Write the number you roll in the box, and color that many
beads. Roll the die again, and do the same on the next set of beads. Circle
the chain with fewer beads.

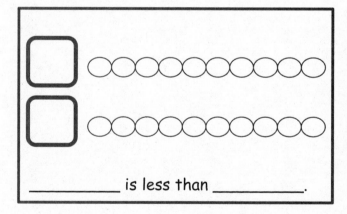

_____ is less than _____.

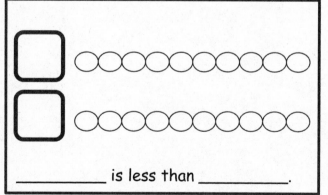

_____ is less than _____.

On the back, make more chains by rolling the die. Write the number you
rolled, and then make a chain with the same number you rolled.

square path letter trains

Write the letters of your first name on the linking cubes.

Connect the cubes.

Draw the cube train you made.

Draw

[blank drawing box]

(Students use a dry erase marker to write on the linking cubes.) Compare your train with a friend's.
Which train is longer? Count the cubes in each train. Which number is more? Which number is less?

Name _____   Date _____

Color the shapes. Count how many of each shape is in the shape robot.
Write the number next to the shape.

Red

Yellow

Green

Orange

Look at the robot. Color the shape that has more.

Are there more or ?

Are there more or 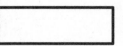 ?

Are there more  or ?

Look at the robot. Color the shape that has fewer.

Are there fewer 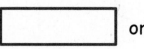 or  ?

Are there fewer or ?

Are there fewer or 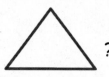 ?

       Lesson 21:   Compare sets informally using *more, less,* and *fewer.*

Name _____   Date _____

Draw a shape to make the sentence true.

There are more _____than   .

There are fewer   than _____.

There are fewer _____than   .

_____

more than, fewer than recording sheet

Pretend your linking cubes are little baskets.

Make a small clay ball for each basket.

Put a ball in each basket.

Draw the balls in the baskets shown.

Draw

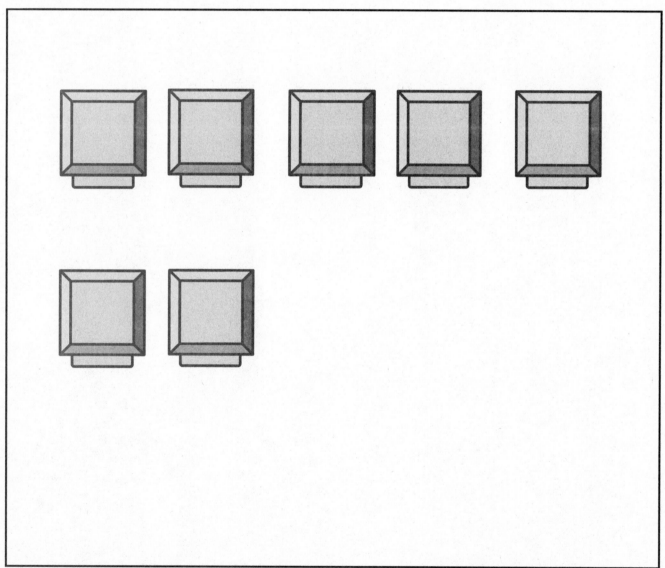

(Each student has 7 linking cubes and a small piece of clay.) Each ball you put in a basket is
1 point. How many points did you score? Are there the same number of balls and baskets?

Name _____   Date _____

Count the objects in the box. Then, draw the same number of circles in the empty box.

Draw a set of objects in the first box. Switch papers with a partner.
Have your partner draw the same number of objects in the next box.

Lesson 22:   Identify and create a set that has the same number of objects.

EUREKA
MATH™

Draw 9 birds.

Draw enough worms so each bird gets one. Now, draw an extra worm.

How many worms are there? Write the number.

**Draw**

Use a ruler to draw a line from each bird to its worm. Did you draw too many worms? How many extra worms are there?

Name _____   Date _____

How many snails? ☐	Draw 1 leaf for every snail and 1 more leaf. How many leaves? ☐
How many pterodactyls? ☐	Draw 1 fish for every pterodactyl and 1 more fish. How many fish? ☐
How many squirrels? ☐	Draw 1 acorn for every squirrel and 1 more acorn. How many acorns? ☐
How many pigs? ☐	Draw 1 piece of corn for every pig and 1 more piece of corn. How many pieces of corn? ☐

Roll the die. Draw the number of dots in the first box. Then, draw a set of objects that has 1 more. Write the number in the box.

Lesson 23:   Reason to identify and make a set that has 1 more.

©2018 Great Minds®. eureka-math.org

Draw 9 birds. One of the birds is not hungry.

Draw just enough worms so that each hungry bird can have one.

How many worms did you draw? Write the number.

 Draw

Use a ruler to draw a line from each hungry bird to its worm. Are there fewer worms than birds?
How many fewer worms are there?

Name _____ Date _____

As you work, use your math words *less than*.

How many kites? ☐	Draw a set of suns that has 1 less. How many suns? ☐
How many hot air balloons? ☐	Draw a set of clouds that has 1 less. How many clouds? ☐
How many octopi? ☐	Draw a set of sharks that has 1 less. How many sharks? ☐
How many chicks? ☐	Draw a set of worms that has 1 less. How many worms? ☐

Roll the die. Draw the number of dots in the first box. Then, make a set of objects that has 1 less. Write the number in the box.

Lesson 24:   Reason to identify and make a set that has 1 less.

Put 10 pennies in rows as shown.

Put one linking cube on top of each penny.

For each cube you put on a penny, draw a square below.

 Draw

(Students have 10 pennies and 8 linking cubes.) Are there enough cubes to cover each penny?
Are there more pennies or cubes?

Lesson 25:   Match and count to compare a number of objects. State which   **115**
                   quantity is more.

©2018 Great Minds®. eureka-math.org

Name _____   Date _____

Count the objects in each line. Write how many in the box. Then, fill in the blanks below. Use the words *more than* to compare the numbers.

_____ is more than _____.

_____ is more than _____.

_____ is more than _____.

EUREKA MATH™

Lesson 25:   Match and count to compare a number of objects. State which quantity is more.

117

Roll a die, and draw a set of objects to match the number rolled. Write the number in the box. Roll the die again, and do the same in the next box. Use the words *more than* to compare the numbers.

_____ is more than _____.

_____ is more than _____.

_____ is more than _____.

Lesson 25:   Match and count to compare a number of objects. State which quantity is more.

©2018 Great Minds®. eureka-math.org

Draw a happy face in a row or line for each person sitting at your table.

Now, draw each pencil at your table in a row or line.

Draw lines to match each happy face to one pencil. Remember, each one gets only one partner!

Are there more pencils or people? Show your work to your partner.

Lesson 26:   Match and count to compare two sets of objects. State which quantity is less.   **119**

©2018 Great Minds®. eureka-math.org

Name _____   Date _____

Count the objects in each line. Write how many in the box. Then, fill in the blanks below. Say your words *less than* out loud as you work.

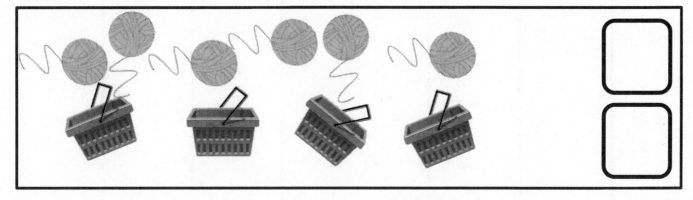

_____ is less than _____.

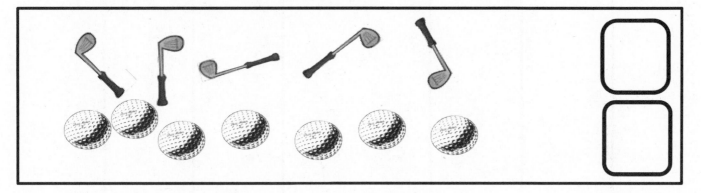

_____ is less than _____.

_____ is less than _____.

Lesson 26:   Match and count to compare two sets of objects. State which quantity is less.

121

©2018 Great Minds®. eureka-math.org

Roll a die, and draw the number of dots in the box. Then, draw a set of objects to match the number. Roll the die again, and do the same in the next box.

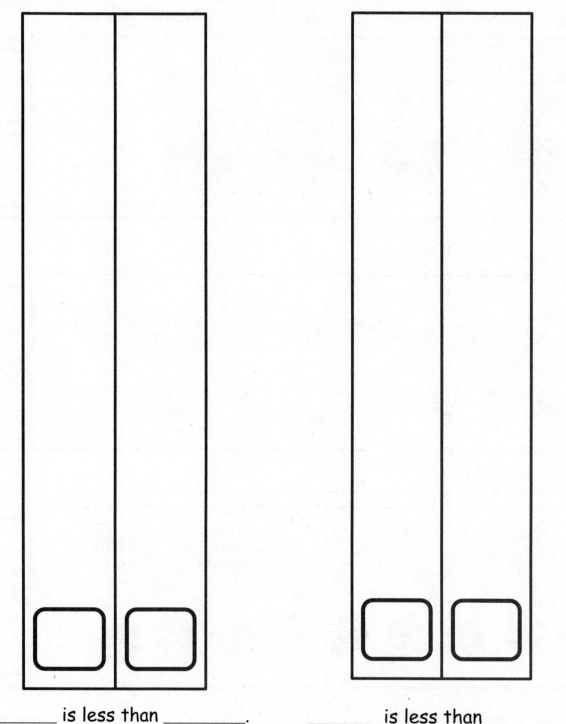

_____ is less than _____.   _____ is less than _____.

Lesson 26:   Match and count to compare two sets of objects. State which quantity
                              is less.

Take one handful of blocks.

Draw all of your blocks.

 Draw

(Provide buckets of pattern blocks for pairs of students.) Compare your handful of blocks with your friend's. Who has more blocks? How did you know who has more or less?

Name _____   Date _____

Draw a tower with more cubes.	Draw a train with fewer cubes.	Draw a tower with more cubes.
___ is more than ___.	___is less than ___.	___ is more than ___.

Draw a train. Draw another train with fewer cubes.

_____is more than _____. _____ is less than _____.

Draw 4 snowmen.

With your clay, make little hats and put them on the snowmen.

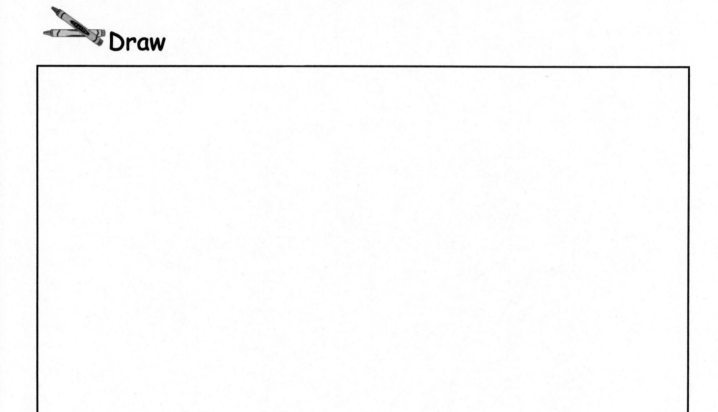 Draw

Now, make 2 more hats for the snowmen that melted yesterday. How many snowmen did you draw? Write the number. How many hats did you make? Write the number. Which number is greater? Which number is less?

Name _____   Date _____

Visualize the number in Set A and Set B. Write the number in the sentences.

3		5
Set A		Set B

_____ is more than _____.

_____ is less than _____.

7		6
Set A		Set B

_____ is more than _____.

_____ is less than _____.

8

Set A

6

Set B

_____ is more than _____.

_____ is less than _____.

9

Set A

10

Set B

_____ is more than _____.

_____ is less than _____.

Roll a die twice, and write both numbers on the back. Circle the number that is more than the other.

Look at the carton of orange juice shown.

Imagine the juice being poured from the carton to the cup.

Draw what the juice would look like in the cup.

 **Draw**

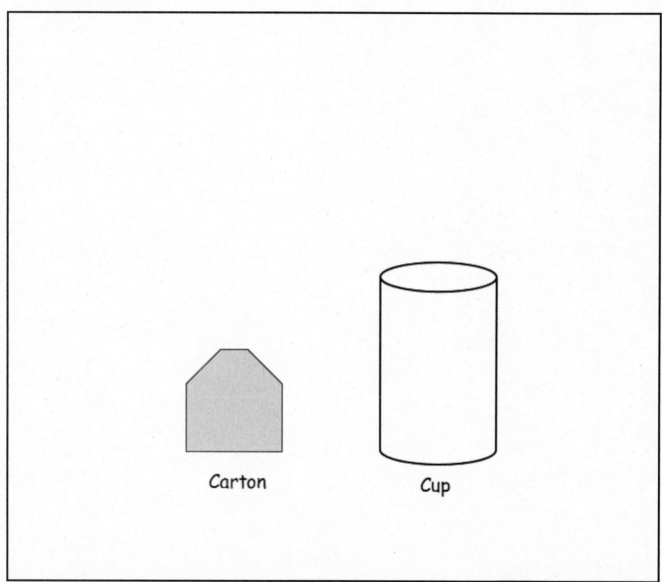

Carton   Cup

Is there more, less, or the same amount of juice when it is poured from the carton to the cup? Look at your friend's cup. Did either of you draw the juice to the top of the cup? Explain to each other why or why not.

Name _____   Date _____

My Capacity Museum!

my capacity museum recording sheet

Lesson 29:   Observe cups of colored water of equal volume poured into a variety of container shapes.

135

©2018 Great Minds®. eureka-math.org

Imagine one big ball of clay and four small balls of clay.

The big ball of clay is as heavy as the four small balls of clay.

Draw the balls of clay on the balance scale that would show they are the same weight.

Draw

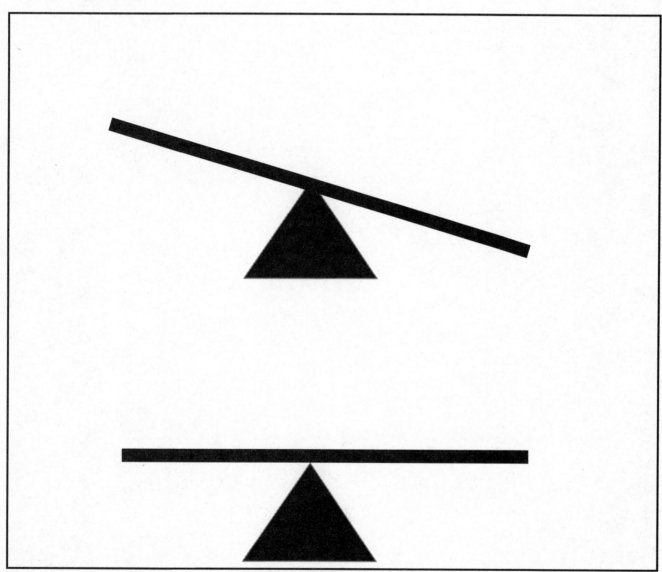

(Circulate as students work to ensure understanding of the task and to see if students are creating reasonably sized balls for the comparison.) Which balance scale did you draw the balls on?

Name _____   Date _____

Clay Shapes

clay shapes recording sheet

Put beads on your string so the beaded part is as long as your hand.

Draw your string of beads.

Draw

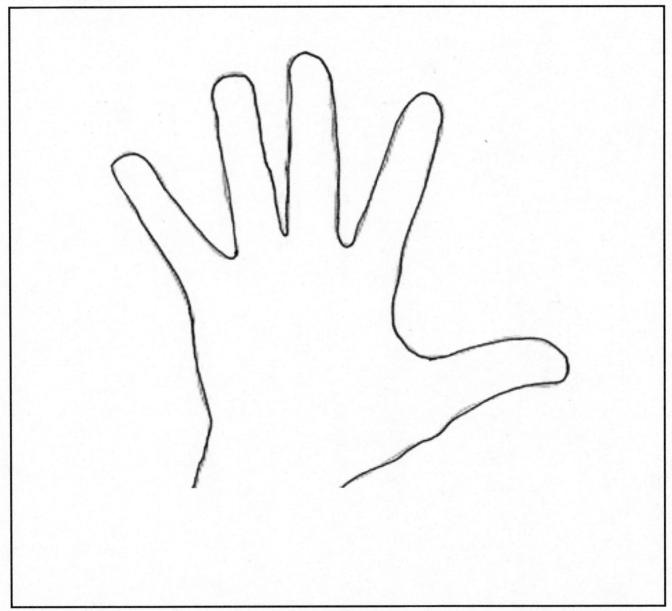

(Prepare elastic strings with a bead tied to one end for each student.) Compare your string to a partner's. Are they the same length? How many beads did you use? Tie the ends of your string together to make a bracelet.

   Lesson 31:   Use benchmarks to create and compare rectangles of different lengths   **141**
                                        to make a city.

©2018 Great Minds®. eureka-math.org

Name _____   Date _____

Listen to the directions, and draw the imaginary animal inside the box.

Draw a rectangle body as long as a 5-stick.
Draw 4 rectangle legs each as long as your thumb.
Draw a circle for a head as wide as your pinky.
Draw a line for a tail shorter than your pencil.
Draw in eyes, a nose, and a mouth.

```
┌─────────────────────────────────────────────┐
│                                             │
│              Imaginary Animal               │
│                                             │
│                                             │
│                                             │
│                                             │
│                                             │
│                                             │
│                                             │
│                                             │
│                                             │
└─────────────────────────────────────────────┘
```

EUREKA
MATH™

Lesson 31:   Use benchmarks to create and compare rectangles of different lengths to make a city.

143

Name _____   Date _____

comparing attributes recording sheet

Lesson 32:   Culminating task—describe measurable attributes of single objects.

145

Credits

Great Minds® has made every effort to obtain permission for the reprinting of all copyrighted material. If any owner of copyrighted material is not acknowledged herein, please contact Great Minds for proper acknowledgment in all future editions and reprints of this module.